baby ORCAS

KIM THOMPSON

CREATIVE EDUCATION • CREATIVE PAPERBACKS

CONT

ENTS

I AM A CALF.

I am a baby orca.

Orcas are also called killer whales.

My mom has just one calf at a time.
I swim in the ocean beside my mom.

I drink my mom's milk. It has lots of fat. It helps me grow blubber.

I swim all around the world! I live with many family members. We are a pod.

We use echolocation to find each other.

I learn to hunt.

I **catch fish**, seals, penguins, and other animals. My teeth rip and tear.

I work together with my pod to hunt.

SPEAK AND LISTEN

Can you speak like a calf?

Baby orcas squeak and click.

Listen to these sounds:

https://www.youtube.com/watch?v=GB5u-vrUGmuA

ORCA WORDS

blubber: a thick layer of fat that helps a whale's body stay warm in cold waters

echolocation: a way to find something by making sounds and listening to them bounce back off of objects

pod: a group of whales that live together

whales: large mammals that live in the ocean

READING CORNER

Chanez, Katie. *Killer Whales (World of Whales)*. Minneapolis, Minn.: Jump!, Inc., 2023.

McConnell, Cathleen. *Toothed or Baleen? A Whale Compare and Contrast Book*. Mount Pleasant, S. Carolina: Arbordale Publishing, 2024.

Riggs, Kate. *Killer Whales (Amazing Animals)*. Mankato, Minn.: Creative Paperbacks, 2021.

INDEX

PUBLISHED BY CREATIVE EDUCATION AND CREATIVE PAPERBACKS
P.O. Box 227, Mankato, Minnesota 56002
Creative Education and Creative Paperbacks are imprints of The Creative Company
www.thecreativecompany.us

LIBRARY OF CONGRESS CATALOGING-IN-PUBLICATION DATA
Names: Thompson, Kim, 1970- author
Title: Baby orcas / Kim Thompson.
Description: Mankato, Minnesota : Creative Education and Creative Paperbacks, [2026] | Series: Starting out | Includes bibliographical references and index. | Audience term: juvenile | Audience: Ages 4-7 Creative Education and Creative Paperbacks | Audience: Grades K-1 Creative Education and Creative Paperbacks | Summary: "Introduce beginning readers to the world of baby orcas with this life science starter. Includes photos, a labeled animal diagram, "Make a Noise" section, glossary, and further resources"-- Provided by publisher.
Identifiers: LCCN 2024043245 (print) | LCCN 2024043246 (ebook) | ISBN 9798889897484 library binding | ISBN 9781682778340 paperback | ISBN 9798889897613 ebook
Subjects: LCSH: Killer whale--Infancy--Juvenile literature
Classification: LCC QL737.C432 T476 2026 (print) | LCC QL737.C432 (ebook) | DDC 599.53/6139--dc23/eng/20250107
LC record available at https://lccn.loc.gov/2024043245
LC ebook record available at https://lccn.loc.gov/2024043246

DESIGN AND PRODUCTION
Design by Rhea Magaro
Production by Beeline Media and Design, Inc.
Art direction by Tom Morgan

PHOTOGRAPHS by Alamy Stock Photo/Juniors Bildarchiv / R304, 12; Dreamstime/Bennymarty, 11, Slowmotiongli, 6-7, 7; Getty Images/GeoStock, 10-11; Shutterstock/Alexander Baumann, cover, Bohbeh, 13, Christian Musat, 4, Happy Whale, 2-3, Mike Price, 5, Pete Niesen, 14, slowmotiongli, 8, Willyam Bradberry, 9

Printed in India